CHEMISTRY
Strategic Paths to Understanding

CHEMISTRY
Strategic Paths to Understanding

Robert G. Bryant

Commonwealth Professor of Chemistry, Emeritus
University of Virginia

iUniverse LLC
Bloomington

CHEMISTRY, STRATEGIC PATHS TO UNDERSTANDING

iUniverse books may be ordered through booksellers or by contacting:

iUniverse LLC
1663 Liberty Drive
Bloomington, IN 47403
www.iuniverse.com
1-800-Authors (1-800-288-4677)

ISBN: 978-1-4917-0328-1 (sc)
ISBN: 978-1-4917-0329-8 (ebk)

Printed in the United States of America

iUniverse rev. date: 08/15/2013

INTRODUCTION

This is an essay on how to study effectively for science courses at the college level. Some of the concepts involved are useful in any course; some are more specific to chemistry, the subject I have taught for over 47 years. I am prompted to write this perspective because my experience suggests that there is a need. I have talked with many students who have worked very hard but have been frustrated when their substantial investment is not reflected in their grade, and more importantly, in their understanding that is necessary to progress in the sciences. This summary derives from my collective experience with these students as well as those who have appeared to be much more successful.

Webster's 3rd Edition of the New College Dictionary defines "study" as

1) *the act or process of applying the mind so as to acquire knowledge or understand, as by reading, investigating, etc.*
2) *To get or perceive the meaning of, know or grasp what is meant by, comprehend, to know thoroughly, grasp or perceive clearly and fully the nature, character, functioning, etc. of* _____.

The same dictionary defines understanding as

1) *To get or perceive the meaning of, know or grasp what is meant by, comprehend.*
2) *to know thoroughly, grasp or perceive clearly and fully the nature, character, functioning, etc. of . . .*

As a student, the challenge is to translate these definitions into useful habits and personal strategies that produce long and short-term benefits. Understanding the target is crucial for success in hitting it.

WHY ARE YOU HERE?

To understand the target, you should first ask yourself, "Why am I here?" "Why am I in this course?" "Why am I in college?" "Why does it matter?" Although it seems too late to ask these questions because you are already here, it is a critical question to examine because the answer can have a major effect on how you approach a subject, how you prepare for examinations, and your short-term and long-term success. Some answers might be:

Because it is required.
Because I need the credits.
So I can get an A.
So I can get into medical school.

Yes, these may be reasons for being in a particular class, but they do not inform the process of preparation and study in a useful way. I hope that you are not in college because you have nothing better to do. If you are not interested, explore other paths to your future. For example, you might try working in different occupations to get a hands-on view of what may or may not be of interest to you. It is interesting to note that some of the most productive people did not earn a college degree; examples are Bill Gates and Paul Allen (Microsoft), Edwin Land (Polaroid), Steve Jobs and Steve Wozniak (Apple). To get a better idea of why you are here is a good investment of time. The question is not whether you know exactly what you want to do with the rest of your life. Most students beginning college have no idea of what their long-term interests are, what their major may be, or what their vocation may be. This is actually fine, the issue is whether you are passionate about finding answers to these and many other questions by exploring deeply all avenues available. You are not here to get high grades, although high grades are fine; more on grades later. You are not here to party, although there may be some good parties. You are not here just for classes or to listen to lectures, the environment is much richer than that and you want to be sure to take full advantage of the scope and breadth of this rich environment in assembling a practical answer to the questions above. But much of this is an aside; the point is to understand why you are really taking each of your classes.

A more useful answer to this first question is a general one that speaks to the value of your education in the first place. *What are you really doing in school?* In an age of rapid access to information, you are not here to accumulate facts, although you may accumulate information. The real work and benefit of education involves building your intellectual and practical skill set to live fully and be a competitive contributor in an ever changing world. This question and its more general answer raise the issue of what you will finally take away from the course(s) that will enable you to be effective. That means being effective in all kinds of new and different contexts, using tools in your personal intellectual toolbox in creative ways to address problems that trouble humankind in creative ways. Yes, it sounds grand, but I hope that is what you are really doing here. Perhaps more important to you is that thinking from this perspective will significantly impact how you approach education and consequently your success in both the short and long-term. To think of a class as providing tools for your personal toolbox for survival and use *throughout your life* changes how you work through a course. In my experience, thinking from this broader perspective has the immediate benefit of higher grades and gives you greater confidence in approaching life in general.

A key part of doing well in any course is to *accept responsibility* for your own survival, and to accept responsibility for your own decisions, large and small, all along the way. You will need to earn your survival. Ten years from now, there is no value in the statement, "They didn't teach me that." This statement will be interpreted as, "I am lazy, didn't bother to learn that, so I can't help you with that problem." Sounds harsh doesn't it? But the point is that you really do have the responsibility for your own education and nobody else can do it for you. Courses help in the process if you recognize how to take advantage of them, but you have to take command of the learning process in a way that meets your goals and addresses your particular strengths and weaknesses; indeed, you know your strengths and weaknesses better than anyone else.

Beware of false motivations. There are many, grades included! What do I mean by that? Most people think that high grades are the goal. I will certainly not argue that high grades are bad, but I will point out that they may or may not reflect the kind of investment that meets your long-term goals. Ten or twenty years from now, nobody will pay much attention to your grades because that is history. The issue at that time will be, "What can you deliver now?" You may translate that as, "Did you retain anything of value from your educational investment?" If you cannot deliver solutions to real problems, course grades, even A's, are completely meaningless. Thus, in your quest for high grades, keep your real focus on the long-range goal so that you carry away from the course(s) tools of lasting value.

It is also useful in focusing on the long-range goals to understand your motivations for your particular direction in the curriculum. Is it really *your* choice? Are you motivated by social pressure, parental pressure, a sense of competitiveness? Some people choose to do what people perceive as hard things because they can derive some joy from being successful at something challenging or hard. That is fine. But, if you are not really interested in an area for the long-term, perhaps ten or fifteen years from now, it may be more joyful to participate in something you really care about. Thus, as hard as it can be at times, take a moment a few times a year to reflect on your real motivations. It helps a great deal to be emotionally committed to your direction when you dig into the specifics of a course.

THE ROLE OF MEMORY
AND ITS FAILURE

For a variety of reasons, many students think that the tried-and-true approach to learning is to memorize a bunch of recipes for how to work problems, lots of facts that can be regurgitated on a test, and that anything you cannot memorize must not be important. For most college science courses, this approach leads to very disappointing results for both the short-term goal of getting good grades, and the long-term goal of being effective in solving new problems in a variety of areas. You do need to remember a few basic things which become the vocabulary of the subject. In chemistry, this includes such fundamentals as the units of various measurements, standard prefixes such as k, m, for kilo and milli, and the basics of naming elements and compounds including the symbols for the elements; the vocabulary if you will, just like French class. Thus, you will need to know that H_2SO_4 is the chemical formula for sulfuric acid; that H, S, and O stand for hydrogen, sulfur, and oxygen respectively, but this effort is actually minimal in the grand scheme of things. It is simply efficient to have these basics at the ready so you can get on to the more important and interesting issues. So, you do need to remember some material, but not really that much. And in the future, you can easily look it up, even now on your telephone. *The issue is not the information, but what to do with the information.*

Although chemistry involves learning a considerable number of concepts, the real goal is how to apply them in real-world situations. There are conceptual foundations that guide thinking in understanding chemical transformations as well as the details of understanding the energetics of these transformations and their quantitative consequences. Often this translates into solving problems, like those nasty word problems people love or hate from math classes. The difference is that now they are mixed up with chemical concepts, perhaps new concepts. However, the math that you need to solve them is the same, and for first year chemistry, it is just elementary algebra. People complain about the math in these courses, but the problem is not the math at all, it is to know how to apply it in the context of a chemical problem, or an economic problem or . . . . For example, almost all entering students can solve the expression, $x^2-3x+2 = 0$ for x, the real problem is how to get to such an equation from the context of a chemistry *word problem*. You see, this an application of the idea presented above. A high grade in algebra is not meaningful unless you can use that algebra now as a tool to solve other problems, chemistry problems in this case. In a real sense, now is the real test of whether you learned the concepts of algebra or not, just as in years ahead, you will discover whether you learned anything lasting in chemistry.

Back to memory issues. Textbooks today usually provide step-by-step methods for solving problems in general chemistry. Books have evolved in this direction, and publishers now demand this approach because there is the sense that the marketplace requires it as a component of a modern textbook. These examples can be useful; however, they can also be toxic to long-term learning. Why?

Some students decide that the way to do chemistry is to memorize a step-by-step method for solving a problem of some type. Most of these students will not be able to answer a question or solve a problem where something in the problem is changed or turned around. People who memorize a procedure (take this, divide by that, multiply by the other thing) trip badly when the problem is changed and this memorized approach leads them to the wrong answer. The frustration is that such students memorize a great deal and thus invest a good deal only to get the wrong answer. They may be angry with the course and the subject. They may wrongly conclude that they are no good at science anyway and they should abandon their dream of a career in science, engineering, or medicine. They have made a large investment of time studying, but the investment was not effective and did not serve them well in reaching their goals.

You may rightly ask, if no recipe, how do you get to the answer? Well, getting to an answer usually involves a sequence of operations or steps, and retrospectively you could always claim that there is a recipe of sorts for getting to it. In fact, there may be several, some much more efficient than others. The point is that you want to understand the material well enough so that you can *create your own recipe* for getting to the answer. Almost always, there are major efficiencies when you have this level understanding. On the very practical side, computers are very good at solving problems with recipes, i.e., programs that knowledgeable people write; and, computers are much faster at it than humans. Thus, your future value will not be based on executing recipes even if you can recall them. Rather, your future value will be in creating solutions to problems not yet formulated. Your value will be in being able to write the program, not be the computer. Indeed, part of your value will be in the ability to reduce a complex situation to a soluble problem in the first place. To do that kind of work, you need a deeper understanding of your subjects than memorized recipes. Yes, it sounds hard, but once you dig in and understand how to approach the material, it is not so hard, and it is very rewarding, both in the short and long term. In summary, if you find yourself memorizing simple approaches to solving problems, or memorizing vast amounts of material, you are probably going to be frustrated in a major way. The need is to invest your effort differently, and in doing so, go deeper.

Although many instructors provide the relevant formulas for the basic physical relationships on the front page of examinations, it makes some sense to have the basic ones in memory. There are not many, so the effort is small. The reason for having these facts in memory is that they provide the means for thinking about the issues anywhere in a more general context. Later we will use the ideal gas law as an example. The relationship is remarkably simple: $PV=nRT$, where P is pressure, V volume, n the number of moles of gas, R the universal gas constant, and T the absolute temperature. From this relationship a great deal of reasoning may follow, and it is efficient to have this relationship available without needing to look it up each time. However, some people try to memorize all the algebraic rearrangements of this relationship! There is no need to do that if you can take advantage of the algebra that you know to rearrange the equation and "derive" the relationship you need.

I mentioned that often professors give students the formulas as reference material somewhere on the examinations in a course. Why does this make pedagogical sense? As noted above, the issue is not about memorizing formulas, it is about thinking with the relationships to understand what is going on in any situation. The issue is what to do with the relationships to solve problems, rearranging them as needed, deducing conclusions from them for a specific circumstance. The thinking part is the real issue, not memorizing the formula. In the years to come, you will almost always be able to look up the information you may need for problem solving anyway, providing the formulas gives you a realistic perspective for what you may do in the future. The so called open-book examinations can be even better for testing mastery of material. While it sounds great to have an open book examination, students usually find the open-book experience more difficult because all the "useful information" is somewhere in the book, none is selected for you on the front page of the examination. A significant part of the exercise is deciding what you need to use out of the large amount of data present in the book. Of course, ten years from now, that is how you might use the material, so the open-book exercise is quite realistic. What is missing are the hints implicit in the information provided with the problem for closed-book exercises.

Having the basic formulas available in your head also provides, in the long-term, the foundation for thinking about problems. If you become an active scientist, you will often do your creative thinking starting from what is already in your head. Of course, you will then refine the ideas with detail and usually make very quantitative examinations of the consequences, etc. Most of these detailed tests and analyses you will not do in your head. The process reminds me of a friend who was in sales. His comment was that you cannot negotiate to buy an automobile if you cannot do the simple math in your head. The same is somewhat true in the sciences. In discussions with colleagues over puzzling problems, the creative thought process usually draws from the basic foundations of the discipline first, and then the subtle and more complex parts are added later. In summary, while memorizing recipes usually leads to frustration, having the few basic relationships available from memory is usually efficient. More on quantitative preparation later.

PARTICIPATE!

It seems obvious that to study is to participate in learning. But recognizing *real participation* is apparently not so simple. I frequently use the analogy to athletics as an example of what to do or not do in studying for a class. Any athletic endeavor works in this context, but I will use two. An aerobic sport such as swimming requires considerable training in the water in order to develop the strength and stamina for effective competition. Watching swimming practice does not get the spectator into good shape for competition. In order to get into shape, you need to be in the pool, working hard on stroke mechanics as well as aerobic effort. Watching someone else do the work does not get you very far. Similarly for basketball; if you watch a game, it appears simple. People run around, throw the ball to each other, and finally throw it at the basket while the other team tries to stop them. If you don't spend any time on the basketball court yourself, you actually have very little idea of what to do to be an effective player. The clear message from these two examples is that for studying chemistry, you need do the equivalent of getting into the pool or onto the court. You need to do the heavy breathing personally and learn to react under the pressure of the situation to sharpen your skills; to do that you need to participate intimately in the effort.

To "get into the pool" sounds easy, but many people seem to miss the point when it comes to understanding effective exercise for a chemistry course. I expand below on some things that do not work very well for getting in academic shape. The key point here is to *avoid being a spectator*. Spectators are passive. They are not participating in the actual activity, only watching. Watching may be fun, but when it comes to developing effectiveness in the activity, it is not usually a sufficient investment. *The problem is, people often do not recognize passive activities for what they are and spend remarkable amounts of effort doing them to achieve little advantage.*

A common problem is mistaking recognition for understanding. You may recognize some relationship, formula, or concept; perhaps you have seen it in a previous class. The tendency is to pass over it because you have the illusion that you understand the concept because you recognize it. This is particularly a problem in the first chemistry course for students who have had Advanced Placement chemistry experiences. The problem is similar to recognizing that a group is playing basketball and mistaking the recognition for understanding how to play the game well yourself. You may recognize a tune on the radio, but that is a long way from being able to sing or play it well yourself. Recognition of a concept may lead to lack of commitment in working with the concept to fully grasp it; i.e., not diving into the pool for a good workout.

In one sense, sitting in a class can be a passive activity. To maximize your gain from a class or a lecture, you need to *think your way through class*. You need to be asking yourself questions all throughout the presentation. What is the point of that? Why is this emphasized? How does this follow from the last class, etc? You can even ask such a question in class so everyone can benefit. Of course, classes become a very passive activity when you nod off to sleep, text on the telephone to arrange lunch, or look up the latest information about fantasy football on your personal electronics. Economically, you are wasting tuition by these activities.

Beware of an overemphasis on notes. Taking notes can be an aid to concentration, but the process can also be a distraction if you already know well what you are writing down. Being a participant translates into taking notes that matter to you for refreshing your memory about the emphasis in the class, concepts or examples that are not in other materials such as the textbooks, and examples that are particularly interesting. If you take notes, it makes no difference how neat they are as long as you can read them later. Some students invest remarkable amounts of time creating neat notes by recopying them, sometimes with multicolored pens. This is largely a passive activity because your brain is not processing the material very much while you recopy your notes; you are copying them, not creating them in the first place. I recommend not investing time on this whole endeavor, but invest the time in a mentally more active process; i.e., jump into the pool for some intellectual heavy breathing.

If you look at a current textbook, they are very well organized. There are sections, subsections, and specific points of emphasis all set off with different colors, font sizes, and indents. The book outline is abundantly apparent just from turning through a few pages. Thus, don't bother to make an outline of the book because you already have one, and outlining the book is a passive activity. Many, perhaps most, people make extensive use of underlining what are thought to be critical aspects of the text. Often this translates into a passive activity where you may recognize that something may be important, so you mark it to think about it later, and later, and perhaps never. You are not necessarily very actively engaged when you do the underlining. If you do it, make sure you do the thinking that should go with it at the same time.

Many textbooks come with problem books as an adjunct. These books have a large number of worked out problems showing every detail of the calculation or whatever is involved in answering the specific questions. A particularly ineffective use of your time is to read these books one problem after another. Reading worked out problems is largely like watching swimming practice to get into shape. The worked problem example can be very helpful only after you have made a substantial investment yourself in trying to construct your own solution to the question. The struggle puts you into the game, the intellectual swimming pool as it were, and you are actively working with the concepts and processing them even if you get good and stuck. Only after you have a really good investment and still cannot see the key to the issue does it make sense to dig out the problem book to get the clues to the solution. It is kind of like calling time out and talking to the coach for a minute to get back on track. The solution, when you read it after a major investment on your own, will create a burst of understanding that will serve you very well. Fine. Don't read the next problem in the problem book. Rather, put the thing away again some place that is inconvenient so you have to struggle to dig it out again. Making the worked out answers somewhat inaccessible assures that you have done the major part of the intellectual exercise yourself before you call time out and have another talk with the coach. Think of most athletic contests. There is a limit to the number of time-outs that you can call. You cannot always run to the coach. Instead you need to utilize your own problem solving skill so that they improve and you can use them when you need them.

Study groups can be useful just as scientific meetings can be useful and foster discussions of critical issues. However, often study groups can be a major spectator event. If these are problem solving sessions, typically one person in the group does the heavy thinking and demonstrates how to get the answer, the rest of the group watch. Spectators don't learn very much this way. Thus, make sure that you are making a good investment for your time if you participate in extra sessions where someone answers questions or where you work problems in groups. If someone is answering a question in detail that you already understand very well, you are probably wasting your precious time that could be better invested in another activity that engages *you* more directly. If you are going to participate in these kinds of activities, go to them with your list of questions or concerns, and when they are answered, head back to a more efficient investment of your time to meet your goals. A decision to do something else is simply taking command of your own education and making the most of your time. No one should be offended by this.

I will argue that working problems and constructing several approaches to solving them are effective means for developing an understanding of a subject. However, there is a subtle way that this effort can also turn into a spectator activity too. Textbooks and problem books often have a large number of questions you can work on or problems to work out. Many are often similar although the quantities or numbers in them may be different. There is no particular advantage to working the same problem ten times where only a number is different here and there if the fundamental approach is the same. This becomes a boring activity and a waste of time. You drag yourself through the exercise more or less mechanically and your brain goes into neutral. For the most part, you are not learning anything except how often you can make errors punching numbers into your calculator or computer. So, while this may not be quite akin to watching swimming practice, it is not getting you into much better shape either.

WHAT TO DO?

What is actually helpful? There are several answers to this, and some particular aspects will be expanded in the next sections. Some answers are obvious and traditional.

Read the textbook material ahead of class. It is alright if you don't understand everything at the first pass because that gives you specific reference points to focus on in class when the same concepts are addressed, now for the second time. Having a good idea of what is to be presented in class ahead of time gives you the confidence to focus on differences in approach, recognize different perspectives on the same ideas, and raise well-founded questions when you may be missing something.

Participate in class. As noted before, it is easy to lose your focus and drift off into an argument with yourself about what to have for lunch. Active participation in class can help avoid this distraction. You can analyze the logic of the presentation. Determine what is different from the text and what is similar. Notice the emphasis. Some people working only for a high grade will think that this is the most important thing to get from the class because the emphasis in class generally signals the emphasis on examinations. This is usually true. This aspect of the course serves only short-term goals, doing well in class in terms of the grade. It is, indeed, part of the gamesmanship of the academic ritual. It makes some sense to pay attention to this, but far too many people make the short-term goal of a high grade the only goal and miss the long-term goal which is the only one that matters in a decade.

Work problems. Problem solving is the single best investment of your time for growing your understanding and preparing for both long and short-term applications. Problems may involve constructing a strategy for synthesizing a particular molecule or designing a sequence for a chemical separation. Problems may also involve quantitative reasoning and numerical answers from which you may draw conclusions. The next section examines this active process.

PROBLEMS, PROBLEMS, PROBLEMS, AND THE SOLUTION

Solving problems in any subject is an exercise that tests <u>and extends</u> your understanding of the material. Before you start, make sure that you have read the appropriate material and examined any related materials from class sessions. Your preparation may include examining the example problems in the textbook that may be worked out in detail with explanations of each step. Beware! Avoid memorizing a sequence of steps because the sequence is rarely general enough to serve your long-term goals. Some textbooks provide very detailed protocols for solving particular problems that may include filling in a series of boxes or diagrams and finally dividing the content of some box by that of another box. I do not use this approach as a teaching tool in my classes because in my opinion the process hides the underlying ideas and takes the critical thinking process away from the student. Further, this approach encourages memorizing a recipe that you will soon forget. In the long run, this is not an approach that will serve you well; if you forget the diagram, you have no idea how to think through the issues or be creative using them. Your goal is to be able to assemble your own sequence of steps to solve new problems and modify it as needed based on your understanding of the underlying principles.

With this preparation done, now attack the problems, usually provided at the end of the chapters. There is now an implicit problem with the organization of most textbooks; the publishers often group problems by type and section headings in the chapter. For the first pass through the material, this organization appears to be helpful because you readily know where to look to see an appropriate example, perhaps a worked-out one. The problem is that the first step in solving a problem is to *identify what type of problem* it is, and this step is largely removed from the student by the publisher because they already tell you by the organization of the chapter. I will suggest a way to deal with the organization issue later.

Now dig into the first problem and try to solve it yourself. Avoid the temptation to flip immediately to the worked out example in the text. Remember you have put the supplementary worked out problem book far under your bed under a pile of undone laundry or equivalently inconvenient place so there is less temptation there. Think through what the problem is about, what principles are relevant here? Often the approach will be to apply some simple algebra to deduce a quantitative answer, but the real difficulty is recognizing from the word problem, how to rearrange the appropriate relationships to get what you need. Even if you are not getting the answer immediately, your struggle with the problem has you actively engaged with the material. You are processing it, trying ideas out, perhaps discarding a bunch that do not work; but you are *in the pool getting a workout.* If you have done your reading actively, you will usually be able to figure out how to work to an answer and almost all textbooks now provide you with the answer to half if not all of the problems. What the book does not provide, the web does in most cases, so there are few secrets these days.

An aside on the opportunities and pitfalls of the web is in order here. A few years ago I asked an advanced class a question about what to do if your observations in the laboratory came out in a particular startling way. The first answer was "Hop on the computer and 'Google' it." Hmmm. The point I was trying to make was that if you were the scientist who first made the observations, what is the next step? In this case, the answer is not on the web because nobody else knows about it. You cannot be a spectator. You are either a player (on the court) or you are not! It is a good idea to treat the web like the worked out problem book much of the time to avoid excesses in the spectator sport.

You either get an answer or you get stuck. After some time trying but failing to solve the problem, you need a clue to get you started. Now it may be time to check back in the chapter to an example to get an idea that will spark your approach to the problem. Usually, this is all that is required if you are really thinking and are not just memorizing. If after another good effort, at least 10 minutes, you are still stuck, it is OK to climb under the bed, dig through the laundry and retrieve the worked out problem book to see a detailed answer. Read it, then climb back under the bed and put it away again.

Now you have either solved the problem yourself, or solved the problem with help. You are now just getting started, don't stop there, you have just completed the first lap in the "academic practice pool". Now you want to study the problem, i.e., ask a series of questions about the approach, about the mathematics if there is any, about alternative approaches (there are almost always several ways to solve a problem), about related questions, about turning the question around and asking it backwards, etc. By solving the problem, you have opened new insights, but you want to capitalize on these insights before you walk away. The second effort here is at least as important as the first in gaining a useful and long-term understanding of the material.

AN EXAMPLE

I will use something most people study in high school, namely, the ideal gas law. The ideal gas law is remarkably compact and also very useful in a large variety of contexts. It is usually summarized by the equation

$$PV = nRT \hspace{4cm} [1]$$

where P is the pressure, V the volume, n the number of moles of gas (independent of what the gases may be), R the universal gas constant, and T the absolute temperature. There are about 25 different kinds of problems that may be addressed by this relationship. You can try to memorize 25 different recipes, or you can take advantage of all the tools you already have and derive them as you need them. To do this, you need to understand the real content of this powerful summary of gas behavior under typical conditions of temperature and pressure. The tools you have are largely those of elementary algebra and mathematics so the idea is to use these outside of math class to develop a sound understanding of this gas law and other relationships. Of course, you can test your thinking against practical experience as well!

As a first step, this ideal gas relation contains 4 variables, P, V, n, and T. R is a constant which if we cast it in convenient units may be written as 0.08314 L*bar/mole*K. Here I use units of pressure in bar rather than atmospheres, but for our purposes, this is a trivial detail. Or is it a trivial detail? How do you go from one value of R to another? This is a good question, one that a person studying this relationship should work on to understand. It is an easy transformation with the relation 1 bar is 10^5 Pa, and 1 atm is 1.01325 *10^5 Pa. So

$$0.08314 \frac{L*bar}{mol*K} * \frac{10^5\,Pa}{bar} \frac{atm}{1.01325*10^5\,Pa} = 0.0821 \frac{L*atm}{mol*K}$$

This is already an example of how useful it is to keep track of units in addressing quantitative problems. Notice that you cannot make the error of dividing by 10^5 if you make sure that the units cancel correctly, i.e., the factor 10^5Pa/bar cancels the units of bar in the original value of R, so that the first factor converts the pressure units to Pascal, the second factor cancels the units Pa, and leaves atm in the numerator so that the conversion of R from one set of units to the other is complete. I strongly recommend that you keep track of units to avoid silly mistakes. If the units or dimensions are not correct, your answer will not be correct either.

Back to our 4 variables in Eq. 1. The basic idea here is that given any three, you can always solve for the remaining one. All you have to do is rearrange the equation to solve for the variable you don't know. Example, rearrange to solve for the pressure:

$$P = \frac{n*R*T}{V} \tag{2}$$

If you know n, T, and V, you get P by straight-forward substitution, sometimes referred to as plugging and chugging, or *plug and chug*. The *plug and chug* work is easy, the issue is understanding how to get the problem to this stage.

A specific example: Given one mole of air at 25°C contained in 1.00 L, what is the pressure? Three of the four variables are given directly, so this is a straight forward *plug and chug* with the rearrangement of the ideal gas law in Eq. 2.

$$P = 1.00\ mole * 0.08314\ L*bar*mol^{-1}K^{-1} * \frac{273.15 + 25.0}{1.00 L} = 24.8\,bar$$

Usually the problem is given to you in a more practical setting and the equation rearrangement is not immediately apparent. An example shows the issues.

Problem: Assume that you have an automobile tire running at 30 psi on a day when the outside temperature is 25 C, and you want to know what the pressure change is when the temperature of the tire increases by 30 C. Using Celsius temperature here; could use Fahrenheit to complicate the problem with temperature conversions or other forms of the ideal gas law. You might think about that alternative. Before we finish the problem definition, we can rearrange Eq. [1] to think about what the issues are here. Note that this is a practical problem because if your automobile tire suffers too large a pressure change, it may fail; not a good thing! Also, I am using English units for pressure, pounds per square inch, which is what tire gauges use. Be careful, there is a subtle issue here. Tire gauges are in that class of pressure measuring devices that work against the atmospheric pressure and measure the excess pressure over 1 atm. Thus, the actual pressure in the tire is atmospheric pressure, 14.7 psi, plus the pressure gauge reading.

$$P = \frac{nR}{V} * T$$

[3]

I have written the equation so that nR/V is separated from the temperature. We can now write a pressure for each temperature.

$$P_{55} = \frac{nR}{V} * (273 + 25 + 30)$$

[4]

$$P_{25} = \frac{nR}{V} * (273 + 25)$$

[5]

Here I have converted the Celsius temperatures to temperatures on the Kelvin scale with the relationship that the Kelvin temperature is 273.15 plus the Celsius temperature. To simplify I have rounded to 273. The units of temperature are now K. Why is this so convenient?

Now we may subtract the equations [4] and [5] to derive the pressure difference caused by a temperature difference.

$$P_{55} - P_{25} = \left(\frac{nR}{V}\right)(30\ K)$$

[6]

This appears to be almost at the level of *plug and chug*, but there is a problem. We need to know the quantity in the parentheses. There are several ways to get n and V or their ratio. Indeed, if we know the initial pressure, which very likely we do or can easily measure it, then Eq. 2 provides the ratio nR/V. In this case, we know that the initial pressure is 30 psi as measured by a tire gauge. So the pressure inside the tire is 30 psi in excess of atmospheric pressure or approximately 14.7 psi + 30 psi, or 44.7 psi. Now from Eq. 2, we may solve for the value of (nR/V) which is (44.7 psi/(273 + 25) K or 0.15 psi/K. Hmmm, so now multiplying by a temperature gives a pressure, so the units work out fine. Notice, we did not need to know either the volume or the value of n to get to this point. We can now solve for the pressure change by substitution.

$$P_{55} - P_{25} = \frac{0.15\ psi}{K}(30\ K) = 4.5\ psi$$

[7]

We have an answer. Students usually check that this is a correct answer somehow, often by looking in the back of the book for the list of answers, then race on to the next problem. But having solved the problem is only the first stage of the workout; stay in the pool and keep swimming. The workout can be very efficient at this point. So what do you do?

Once you have solved a problem, it is usually very valuable to go back through the problem to analyze the steps used to get a solution. In this practical but somewhat complex problem, we had several embedded problems, a common issue in practical applications. A trivial one was the temperature conversion. Notice that the same problem could be given in a different temperature scale. What can you do about that? There are several approaches probably the simplest is to convert to the Kelvin scale before you do anything else, but think for a moment about the ideal gas law at the level of Eq. [1]. Notice that you could transform the equation to use a different temperature scale such as Fahrenheit, but the equation would look different because the zero of temperature now corresponds to a negative value on the Fahrenheit temperature scale. It is probably inefficient to recast the gas law to use Fahrenheit temperature given how simple it is to convert to the Kelvin scale before using the form of Eq. [1], but a thorough understanding of the gas laws would include understanding what you would do to recast Eq. [1] for a different temperature scale. Note that it is just the application of simple algebra here to do that; nothing you don't already know is required. Doing something like this puts you in the pool getting into better shape.

A second sub-problem was figuring out the coefficient (nR/V). Ask how many ways could you do that? Of course you could be given n, and V; then it is *plug and chug* to get this factor. Think about n for a minute. Air is a mixture of gases. How do you get n for a mixture of gases? This question points to a fundamental aspect of the ideality assumption in the ideal gas law. All gases contribute the same regardless of their identity, regardless of their molecular mass. Lots more buried in this statement, but we will not teach the course here, but focus on study approaches. So you might do well to ask, how can I deduce n if the system is a mixture of gases? If you knew the masses of each of the gases you pumped into the tire, you could figure out the number of moles of each and add them up. To do that you take the mass and divide the mass per mole of the gas and that gives you the number of moles. Fine. Air is practically a constant mixture of gases, largely about 20% oxygen and 80% nitrogen. A critical feature, of course, is the variable contribution of water, which dominates the changes in the barometric pressures we hear about all the time. Hmm, wet air is lighter, right? The point here is that if you deduce the average molecular mass of air, which is just a sum of contributions weighted by the fractions noted above, you can get the number of moles from the mass of air delivered too. Interesting idea, average molecular mass of a mixture; it is practical sometimes.

Back to the factor (nR/V). How did we actually do it? We recognized that if we know P and T, then the quantity (nR/V) is fully determined and it is a simple plug/chug problem to find the value needed. But now ask, is there a time when I would not even need to know this factor? The problem asked for a pressure difference. What if it asked for a per cent change in the pressure, which is essentially a pressure ratio? In this case, taking the ratio of Eq. [4] and [5] gives the pressure ratio in terms of a temperature ratio; the factors (nR/V) divide out. So a variation on this problem which looks essentially the same has a different twist here that you can anticipate if you continue thinking through the problem after you solve it.

Examine the factor (nR/V) further. Notice that if we write it as (n/V)*R it is simply a gas density multiplied by a constant. Moles is an amount of material. If we divide an amount of material by a volume we have a density. In this case, the units are moles per unit volume or moles/L if we use liters as the volume unit. Wow, moles/L is also the molar concentration of the gas, which is hiding in this expression! So if someone gave the problem with the gas concentration provided in moles/L that would provide a direct handle on the coefficient (nR/V). So, you have uncovered another variation on the issue and the benefit is a deeper understanding of the original relationship. To carry this further, just for fun, see that the original gas law may be rearranged

$$P = \left(\frac{n}{V}\right) * R * T.$$

So the key point in this rearrangement is that the pressure is a measure of molar concentration multiplied by R*T. At constant temperature, pressure is directly proportional to concentration, and can be a very useful measure of concentration if you know T.

Scientists love pictures. Very often this statement may be translated to scientists love graphs to represent how one variable responds to others. In this case, the pressure in the tire changes because of a change in temperature. If you were consulting for a tire company and wanted to provide a graph of how the tire pressure responded to changes in outside temperature, what would you do to compute it? Well, we derived a relationship between the change in pressure and the change in temperature, Eq. [6]. We may make a graph of that result putting the pressure change on the vertical axis and the temperature change on the horizontal axis. The result will be a straight line, which is shown below for n =1 mole and V=10 L.

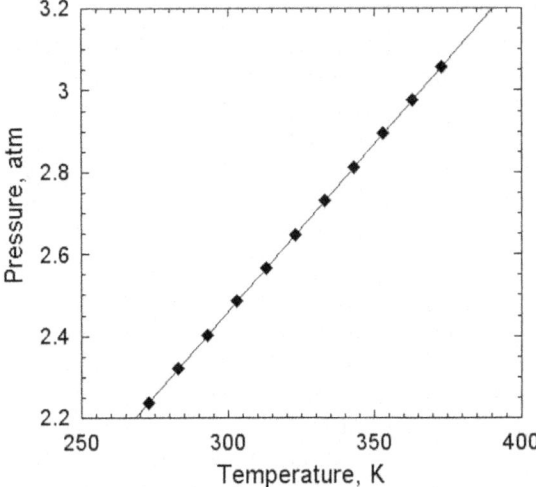

The slope is (nR/V). That is critical! The larger n/V, the steeper the slope and the larger the danger of the tire bursting. So, what determines the quantity n/V? That is determined by how much gas you pump initially into the tire, if the volume is assumed to be approximately constant; but that is determined by the initial pressure, no?

All right, maybe we have stayed in the study of the problem a bit long and need to stop to catch some air. The point is that the work **after** you get to an answer is often very rewarding in both understanding and anticipating variations on the problem that may appear to you in a practical context years later or perhaps on the next examination. In this context you may ask yourself some more simple questions: What "real-world" applications can I think of for these concepts? What other parts of the course may be related to this material? How can I combine this kind of question with other material in the course? What other courses may use this material, e.g., biology, physics, earth science, environmental science, physiology, medicine, among others? If I were making a test, what kinds of questions could I create from these concepts? Here are some examples.

We already looked at the automobile tire example, but there are many related to this one. If you have a home with a well for water, the water is pumped into a tank from which you draw water. In fact it easily gets to the second floor because pumping water into the storage tank compresses the air originally filling it. So, what is the best way to fill the tank to get the most water out before the pressure drops too far?

Here is an interesting challenge. A student bets you $100 that you cannot drink a glass of water set on the ground from a fourth floor window through a straw. Should you take the bet? Why?

Gases are produced by vaporization of liquids and by chemical reactions. Sometimes the result is an explosion. Liquid nitrogen is relatively easy to come by these days because it is used to store biological samples, freeze skin lesions, and many other uses. If an open glass container is placed in liquid nitrogen, gas from the air will condense into the container, mostly liquid oxygen, but nitrogen as well. If the container is closed with a cap or stopper and removed from the cold bath, you have a severe explosion hazard. Why? What is the volume change expected when a liquid vaporizes. Say you have a milliliter or a gram of liquid oxygen in the container, what volume will this turn into when it vaporizes? You can buy dry ice easily. If you put a chip of dry ice in a closed container and it sublimes to the gas, you have a similar explosion hazard. What is the volume change for this case, say for 5 g of $CO_2(s)$?

All combustion reactions generate water, carbon dioxide, perhaps carbon monoxide, all gases. Often the reactant is a solid and the gas volumes created are similarly large. How much paper would you have to burn in order to fill a room with the products of the combustion reaction?

The examples above are questions to be sure, but they are incomplete just the way questions are when you will see them once you get beyond the course. So a significant part of answering these is to complete the question set up. In the last example, you need to recognize that paper is modeled reasonably well as a carbohydrate, so sugar might be an OK starting point, $C_6H_{12}O_6$ for the reactant side. You need to pick a room size and a pressure; no doubt 1 atm is a good start. The interesting thing about this kind of problem is that you are clearly in the intellectual swimming pool now. In fact, you are using the material to think about questions that may be of interest. The material has now become a tool you are using to answer questions that arise in completely different contexts from the immediate course. This capability is a major long-term goal that will serve you very well in many different contexts.

I cannot claim that I have the global solution for success in the sciences, but the approaches described here can make a significant difference. I remember distinctly a series of conversations with a young woman from South Africa who had been trained in the British system. She had been working very hard and doing poorly by her standards, middle to low C. She was doing many of the things that wasted her time and did not increase her understanding. She had beautiful notes, she had outlined the book, she had spent hours reading worked out problems. Nevertheless, the results were frustrating. I convinced her to abandon what was not working with the argument that she had nothing to lose. It took some convincing, indeed, but she tried it. After the next examination I was sitting in my office talking to a handful of concerned students, and she stopped at the door. Tentatively, she asked, "I was wondering what I may have gotten on the examination." I knew because I too was very interested in the experiment. Although I never discuss particular grades in public, I violated my usual practice and possibly the law, to tell her that she had written a perfect exam. The smile that grew is bested only by those of my grandchildren. She stood several inches taller walking out, but the best news is that the dramatically improved results continued unabated for the rest of the year, at least.

ONE MORE EXAMPLE

The balanced chemical equation and the concepts embedded in it are very fundamental to understanding a great deal of chemistry and its applications in many fields. It provides a good example of looking beyond the obvious to develop understanding. Take an example like the combustion of methane, CH_4 in air, which supplies the oxygen.

$$CH_4 \text{ (g)} + 2O_2\text{(g)} \rightarrow CO_2\text{(g)} + 2H_2O\text{(g)} + \text{heat or } \Delta H \qquad [8]$$

This equation is balanced which means that there is the same number of atoms on both sides of the equation, i.e., the same number of C's, O's, and H's on the left as on the right side of the arrow. The equation summarizes the mass relationships in the reaction and indeed, the masses of the reactants must add up to the masses of the products because no atoms are created or destroyed.

If you introduce the concept of a mole, i.e., the mass of Avogadro's number of molecules or atoms, then it says that one mole of methane reacts with 2 moles of oxygen to make one mole of carbon dioxide and 2 moles of water. (The law of conservation of mass in chemistry is valid to the level of at least 10 significant digits, i.e., beyond the precision and accuracy of practical measurements in the laboratory; hmmm . . . , that is an interesting discussion and requires understanding the implications of Einstein's relation, $E = mc^2$, but that is not the point I am trying to make now. It is a point that you may fruitfully revisit when taking physics where the Einstein relation is discussed in more detail. Making this connection later will put your understanding of chemistry into a new and larger perspective that could be very useful for your long-term goals.)

All right, it is relatively easy to balance the chemical equation, and most students come to a college chemistry course with experience in doing this essential skill. But now, how do the mass relationships really enter the thinking? That rests somehow on the idea of a specific mass for each atom, or some collection of atoms. How does the concept of atomic mass originate? How do we measure it? How can you convince someone that the idea of atomic or molecular masses has merit? Above we said that the equation implies that there are the same number of atoms on both sides of the equation, masses too, how does mass count molecules? Why not use volume to count molecules; we buy gasoline by volume, a practical, but not really quantitatively sound scheme! Can you make a graph demonstrating the problem with using volume as a measure of the number of molecules in a practical setting?

Now consider the combustion of ethane, C_2H_6. We may write this chemical equation as,

$$C_2H_6(g) + (7/2)O_2(g) \rightarrow 2CO_2(g) + 3H_2O(g) + \text{heat or } \Delta H \text{ [9]}$$

Some people might argue that the fractional coefficient for the oxygen is stupid because you cannot so easily take a half an oxygen molecule, which would be atomic oxygen. Nevertheless, it is very common practice and very useful to use equations like this to define the amount of heat or enthalpy, ΔH, produced in the reaction per mole of the reactant $C_2H_6(g)$. So, when is it a good idea to employ fractional coefficients? Both equations (8) and (9) imply that the more reactant you burn, the more heat you get. There is a linear relationship between the amount of heat and the amount reacted. So, how could you make a picture of this relationship, i.e., a graph? How could you present it with different units, such as moles of reactant, mass of reactant, volume of reactant?

How can you combine the implications of the balanced chemical reaction with the ideas contained in the ideal gas law discussed above? For example, n= (PV/RT) and a good deal follows from this simple rearrangement. And, when in a practical example might there be problems with using a balanced chemical equation and the mass relationships that it implies?

Well, we can go on, but this essay is not supposed to teach the course you are to be taking, only a brief description of how you might fruitfully approach it. You have better things to do than be a spectator of the process here; it is better to get onto the field now and explore for yourself. You might think of this as a beginning playbook, the real game in elsewhere. Surely, you will add your own keys in the near future.

SUMMARY

In approaching the college science course, try to understand the real long-term purpose of your efforts; work creatively from understanding, not rote memory or fixed recipes; seek the basic foundations and connections; develop a life-long learning pattern and feed the hunger for understanding.

APPENDIX

There is, of course, some gamesmanship in actually preparing for and taking an examination. Some people focus only on gamesmanship and sacrifice the long-term benefits of the whole experience. Nevertheless, there are some things you may profitably consider.

Anticipate your test environment, which is different for every course. Consider for a few minutes how you will feel sitting there trying to focus on the test. What are the distractions, how will you know what time it is and how much time is left to finish? A wrist watch may be a useful aid.

It is common to be somewhat nervous and be tripped by a question that you do not immediately understand how to approach. Think about that ahead of time! What are you going to do? The toxic approach is to forget the rest of the exam and work on this nasty problem because you know that you should be able to do it! Most of the time, one problem will not bury your total score, so move on to the rest of the exam which you can do easily and then come back to the nasty problem issue once you have completed as much as you can. You will eventually get it figured out.

I have made the assumption that the first thing you did in preparing is to make sure that you see the conceptual framework that underlies the work. There are many examples, and this essay is not about the course material itself, but a short example illustrates the point. Most general chemistry students learn that atoms decrease in size as you go from left to right in the periodic table along any row. So this fact allows you to answer questions about which atom is larger than another. Fine, but what is the underlying conceptual framework for this observation. Can you explain that? Are there other implications that derive from the same origin? There are many ways to ask whether you have a good grasp of the foundations here.

Think briefly about the grader. Approximately 1% of students forget to put their name on their examination! It is hard to get credit when there is no name or identification on the examination. Try to write so that a reasonable person can read it. Scribble annoys the reader, and any benefit of the doubt is pushed further in doubt.

Beware partial credit. It contributes to your total score, but it usually shows that there is some understanding missing unless, of course, you have made a simple arithmetic error.

In preparing for the exam, one way to recapture the disorganization of problems you may find on a test is to make a copy of the problems at the end of the chapter. Cut with scissors each problem into a separate strip and scramble them all up. Put the strips in a box, and pull them out one at a time. Now your first job is to figure out what kind of problem it is. It is no longer bunched together with all the other problems of the same type. It seems like it should be very simple to identify the kind of problem, but this aspect of problem solving is more difficult than many students think . . . until the examination.

BRYANT BIOGRAPHY

Professor Bryant received an A.B. in Chemistry from Colgate University in 1965, Ph.D. in Chemistry from Stanford University in 1969. He was a professor of chemistry at the University of Minnesota from 1965 to 1984, Dean's Professor of Biophysics at the University of Rochester Medical School with secondary appointments in Radiology, Oncology, and Chemistry from 1984-1992. He joined the faculty at the University of Virginia as Commonwealth Professor of Chemistry in 1992 and became Emeritus in 2012. His teaching responsibilities over the years included inorganic, analytical, physical, and general chemistry. His research activity involved primarily the application of magnetic resonance spectroscopy and imaging to problems in a broad spectrum of chemical and biological problems. He has published over 250 papers to the scientific literature. He and his wife of almost fifty years have three children and eight grandchildren.